Contents

Earth & Space Sciences
AC9S1U02

TARGETING SCIENCE YEAR 1 © PASCAL PRESS ISBN: 9781925726503

Targeting Science Year 1

ISBN: 9781925726503

Published by Pascal Press
PO Box 250
Glebe NSW 2037
www.pascalpress.com.au
contact@pascalpress.com.au

This edition of *Skill Sharpeners: Science* is published by arrangement with Evan-Moor Corporation, USA.

For sale in Australia and New Zealand.

Cover Design: Janice Bowles
Authors: Guadalupe Lopez, Lisa Vitarisi Mathews
Publisher: Lynn Dickinson
Editor Australian edition: Stella Tarakson
Typesetter: Stacey Grainger
Illustrator: Paul Lennon, *www.dreamstime.com.au*
Cover design: Janice Bowles

Printed in China by 1010 International Ltd.

Introduction

Welcome to your Year 1 *Targeting Science* activity book! It is packed with interesting and exciting activities to help you understand and enjoy science at home or school.

Targeting Science has been written to support the Australian Primary Science Curriculum Version 9.0 and is divided between:

- Biological Sciences
- Earth and Space Sciences
- Physical Sciences

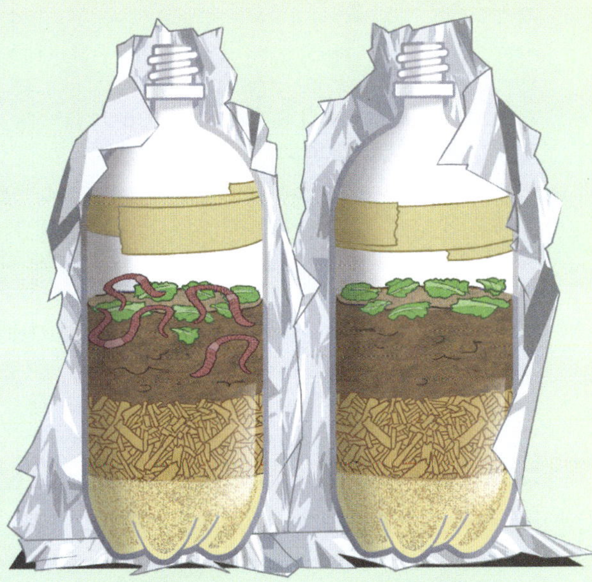

bottle with earthworms   bottle without earthworms

The Science Understanding and Inquiry Skills elements (including Science as a Human Endeavour) are embedded.

Hands-on activities provide opportunities to bring the science concepts to life. The exercises develop process skills such as observing, collecting, recording and organising information and making models of scientific events that happen in the natural world. All activities use easily sourced, inexpensive items.

Topics are introduced with explanations plus images to provide background information. Some lessons include a QR code where you can access a video for further explanations. You will be challenged to match, sort, label, sequence, analyse and answer questions with regular vocabulary practice puzzles throughout the book to help familiarise you with scientific terms. See the Glossary on the next page for some terms you may not already know. Answers are included at the end of the book.

FREE Teaching Guide.

This QR code links to a downloadable PDF of a Teaching Guide to support the material in this student workbook. The guide contains:

- Teaching plans and checklists.
- Graphic organisers.
- Material request forms (to send home for parents).
- Background information about each major topic as well as extension activities.

Use this QR code to access the FREE Teaching Guide.

Front cover – have you looked at the front cover? It depicts what life was like before a significant scientific discovery that led to the development of new technology. What is the difference between Science and Technology? The following quote sums it up nicely.

Science is the process of acquiring knowledge of natural phenomenon along with various reasons. Technology is the application of Science to the solution of problems. (Sismondo, 2018)

Glossary

- **Force** – a push or pull between objects, which may cause one or both objects to change speed and/or direction or change their shape.
- **Investigate** – to perform a scientific process of exploring an idea that requires planning, collecting and interpreting data and forming a conclusion.
- **Meteorologist** – a scientist who studies weather patterns and the atmosphere.
- **Need** – a lack of something useful, required or desired.
- **Objects** – things that can be seen or touched; a material thing that occupies space.
- **Observe** – hearing, seeing or touching something.
- **Predict** - to guess what might happen based on observations.
- **Pull** – to exert force on an object to make it move toward the source of the force.
- **Push** – to press against an object to make it move away from the source of the force.
- **Science** – the process of learning about the natural world by observing and experimenting.
- **Season** – any of the four divisions of the year when the weather changes significantly in temperature, rainfall or daylight.
- **Senses** – any of the five ways to understand and experience one's surroundings. The senses are touch, feel, smell, sight and hearing.
- **Technology** – the application of scientific knowledge for practical purposes.
- **Test** – an observation or experiment that could provide evidence about a scientific idea.

SAFETY
All of the investigations in this book are designed for kids to do safely in the home or classroom. However, adult supervision is recommended.

TARGETING SCIENCE YEAR 1 © PASCAL PRESS ISBN: 9781925726503

Year 1 Australian Science Curriculum Correlations

ACARA Code	Content description	Strand	Sub strand	Pages
AC9S1U01	Identify the basic needs of plants and animals, including air, water, food or shelter, and describe how the places they live meet those needs	Science Understanding	Biological Sciences	2-20
AC9S1U02	Describe daily and seasonal changes in the environment and explore how these changes affect everyday life	Science Understanding	Earth and Space Sciences	21-48
AC9S1U03	Describe pushes and pulls in terms of strength and direction and predict the effect of these forces on objects' motion and shape	Science Understanding	Physical Sciences	49-72
AC9S1H01	Describe how people use science in their daily lives, including using patterns to make scientific predictions	Science as Human Endeavour	Use and Influence of Science	15, 16, 19, 30, 38, 65
AC9S1I01	Pose questions to explore observed simple patterns and relationships and make predictions based on experiences	Science Inquiry	Questioning and Predicting	12, 34-35, 63
AC9S1I02	Suggest and follow safe procedures to investigate questions and test predictions	Science Inquiry	Planning and Conducting	48, 56, 60, 63, 72
AC9S1I03	Make and record observations, including informal measurements, using digital tools as appropriate	Science Inquiry	Planning and Conducting	20, 23, 27, 29, 62-63
AC9S1I04	Sort and order data and information and represent patterns, including with provided tables and visual or physical models	Science Inquiry	Processing, Modelling and Analysing	10, 11, 17, 18, 29, 32, 44, 59
AC9S1I05	Compare observations with predictions and others' observations, consider if investigations are fair and identify further questions with guidance	Science Inquiry	Evaluating	56, 63, 68
AC9S1I06	Write and create texts to communicate observations, findings and ideas, using everyday and scientific vocabulary	Science Inquiry	Communicating	7, 12, 19, 31, 38, 41, 43, 64

Places we find Animals and Plants

In this **science** unit we are going to learn about how plants and animals, including people, meet their needs in the places they live. You probably have some knowledge already about what you need to grow and stay alive.

To begin, let's go on a journey and look for **places where we can find animals and plants.** START here and follow the footprints.

In our schools ...

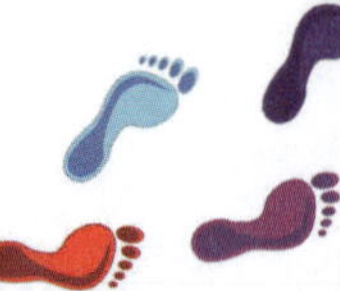

TARGETING SCIENCE YEAR 1 © PASCAL PRESS ISBN: 9781925726503

Places we find Animals and Plants

Food and Shelter

Places and our Needs

We find animals and plants in many different places but only if these places meet their needs. Needs are things plants and animals must have to stay alive. These are **food, water, air, sunshine** and **shelter** which is a place to rest safely. Draw a line to match these animals and plants to the place you think will meet their needs.

Food and Shelter

TARGETING SCIENCE YEAR 1 © PASCAL PRESS ISBN: 9781925726503

Sing this science chant to your child to the tune of "Mary Had a Little Lamb".

Plants and animals
have basic needs,
basic needs,
basic needs.

Plants and animals
have basic needs
to help them
live and grow.

Talk with Your Child

Explain to your child that all living things have basic needs. We call a place that is safe 'shelter'. People need shelter so they can sleep. Plants also need a safe place to grow. Help your child recognise how his or her basic needs are met each day.

What Makes a Place Best for People?

Like all other animals and plants, the best place for us is somewhere that meets our needs. Trace the word under each picture to identify our needs.

If we get these things, we can live and grow!

TARGETING SCIENCE YEAR 1 © PASCAL PRESS ISBN: 9781925726503

What does the puppy need to stay alive and grow? Draw how you would provide for its needs.

What science skill would you use to know if you are meeting its needs?

observation

Animals and Their Needs

Animals that live in the wild must look after themselves. They have to live in a place where they can find the things to meet their own needs. Sometimes in droughts, floods and bushfires, wild animals have to travel to new places to meet their needs. What do you predict would happen if they couldn't find a place to meet their needs?

Look at the picture below. If you were a magpie, do you think you could meet your needs in this place?

Circle parts of the picture that you think will meet a magpie family's needs. (Hint: Magpies eat earthworms, beetles, millipedes, snails, spiders, cicadas, grasshoppers, and small lizards. They drink water, and they shelter and build nests in trees.)

Different Places, Different Needs

There are many types of places on our planet Earth. They may look very different, but the animals and plants that live there are **still able to meet their needs.** All the different places are very important because living things are all very different. People meet their needs differently to fish or lizards.

https://clickv.ie/w/Q0gx

Use this QR code to access a video on this topic.

Look at the pictures below. Put a tick ✔ in the places you will find more people and a dot in the ones where you will find mostly other animals.

Food and Shelter

The Best Place for Me

https://clickv.ie/w/Uxgx

Use this QR code to access a video on this topic.

The place where an animal or plant lives must meet their needs for them to survive and grow. Some of the animals or plants below **do not** live in each place below because it does not meet their needs.

Circle the animals that can live in the home.

orchard

sea

woodlands

TARGETING SCIENCE YEAR 1 © PASCAL PRESS ISBN: 9781925726503

Circle the animals that can live in the home.

Plants and Their Needs

Just like people and other animals, plants have basic needs too. If they don't get these, they will not live and grow.

Imagine you have been given some bean seeds.

What do they need to grow and stay alive? Draw what they need below.

What would happen if the plants' needs were not met?

TARGETING SCIENCE YEAR 1 © PASCAL PRESS ISBN: 9781925726503

Plants are Superheroes

Plants have needs but they also help meet the needs of animals. All animals, including people, depend on plants in some way. They are superheroes because they give food, shelter and fibre that can be used to make things like clothes to keep us warm.

Draw a line to match each animal with the plant that helps meet its needs.

Observe some real animals.

Do you think their needs are being met?

Plants are Superheroes

Plants are essential to meet animals' needs. Can you work out what **need** the plants are meeting in the pictures below? Trace the word below each picture.

Food and Shelter

TARGETING SCIENCE YEAR 1 © PASCAL PRESS ISBN: 9781925726503

Scientists and many other people work hard to learn about and take care of plants. They learn what to **observe** when plants are not healthy and how to meet their needs better. Healthy plants mean a healthy world.

Plant scientists are called botanists

Farmers check their plants' health often

People plant new trees to repair bushland

Horticulturalists look after botanic gardens for us all to enjoy

First Nations Peoples advise us and take care of Country

Gardeners use their science knowledge to grow healthy plants

Observe some real plants. How are their needs being met?

How do First Nations People Care for Living Things?

For over **60,000 years**, First Nations Peoples have lived in Australia. They have always known how important it is to care for the lands, the seas and all things that live there. If you look after Country, Country will look after you!

Here are some ways they did this. Draw a line to a living thing cared for by their actions.

Never cutting down mangroves as they are vital for so many fish, crabs, prawns, birds, turtles and dugongs

Never over-fishing in one place so other animals have food too and our oceans stay healthy

Making controlled burns of small areas so huge bush fires do not destroy plants and animals

Comparing Needs

We know plants and animals all have the same needs, but they don't meet them in the same way. Choose a red and a blue pencil. Circle things that meet your needs in red. Circle things that meet the bird's needs in blue.

Comparing Needs

Put a red circle around the things that meet the needs of the plant. Put a blue circle around the things that meet the needs of the clownfish.

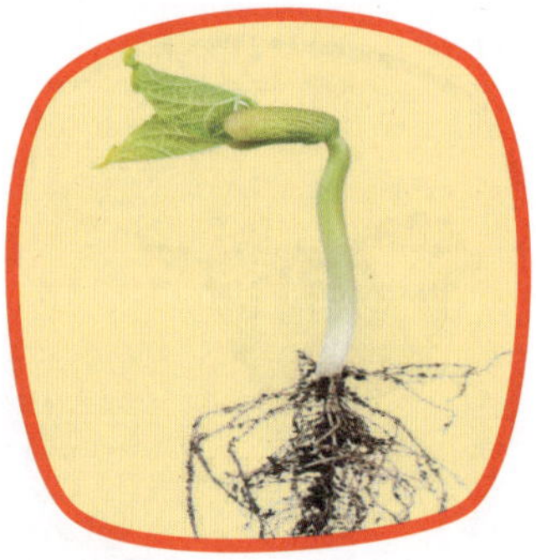

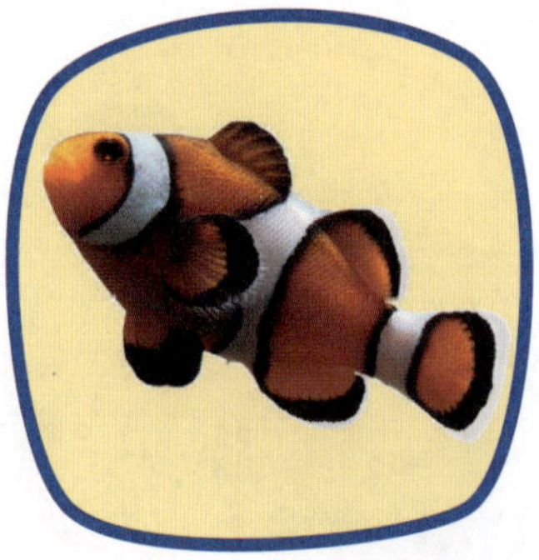

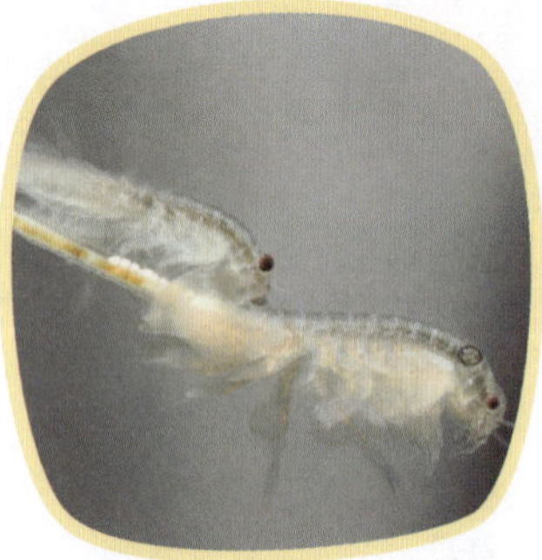

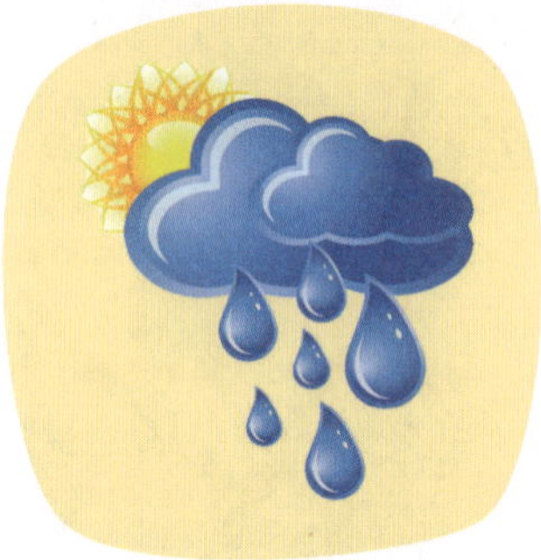

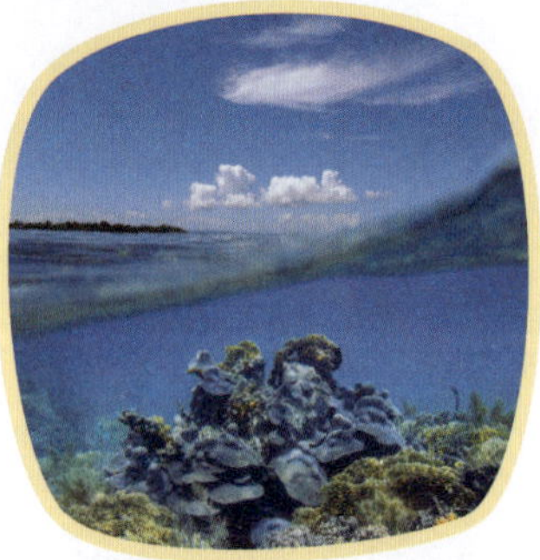

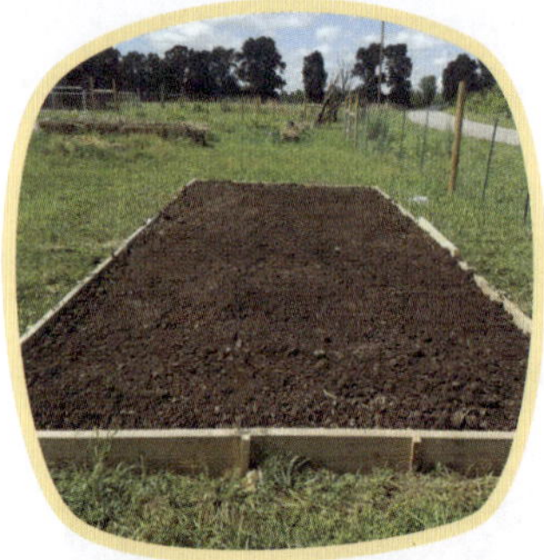

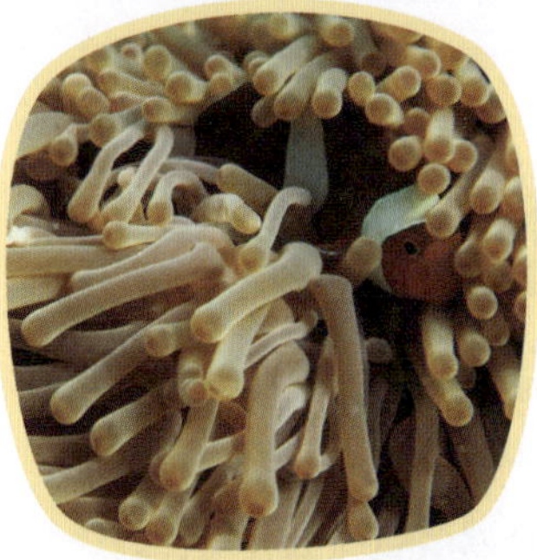

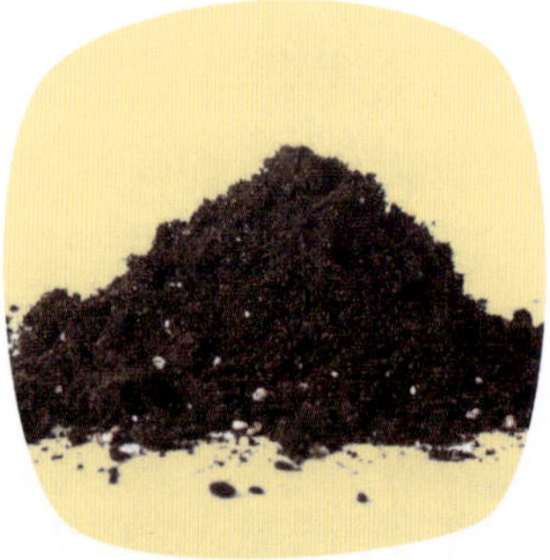

What science skill would you use to know their needs were being met?

observe

TARGETING SCIENCE YEAR 1 © PASCAL PRESS ISBN: 9781925726503

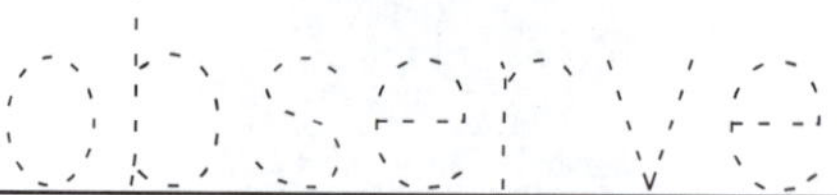

In this unit, we have learnt about the basic needs of plants and animals and how the place they live in must meet their needs to survive.

Write each need under its picture below.

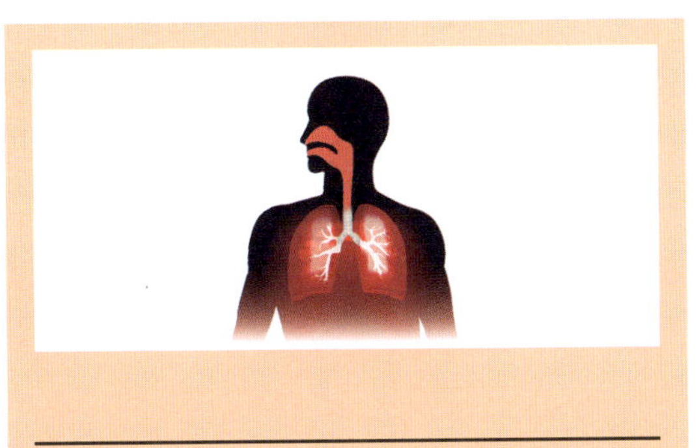

Imagine you find a sick animal or plant. What science skill would you use to know that it is sick?

Which people, including family, friends, people in the community and scientists, could you ask to help better meet the needs of the animal or plant.

Thinking About our Learning

Find a place where you can sit quietly and watch an animal going about its day. It could be a lizard, a bird, an ant or any other animal that is moving around. **Observe** carefully for a while.

Use your science knowledge to draw a picture of how the place is meeting the needs of the animal you have observed.

What would happen if the place where they lived changed and didn't meet their needs anymore?

If possible visit a local petshop, nursery, botanic gardens, farm, zoo, museum with live specimens, keen gardener and talk to someone about how they meet the needs of a plant or animal they take care of.

TARGETING SCIENCE YEAR 1 © PASCAL PRESS ISBN: 9781925726503

Read this science rhyme:

What is **WEATHER?**

What is this thing 'the weather'?
I am hearing what you say.
It's a word we use to talk about
The sky we see each day.

Do we see a blue sky,
With white and fluffy clouds?
Or do we see them dark and grey,
With thunder very loud?

What is Weather?

Do we see the wind
Bending all the trees,
Or the rain drops falling
And dripping from the leaves?

Do we go outside
And feel the sun so hot,
That when we run around
We start to sweat a lot?

Or do we need a jumper
And see the frost beneath
our feet,
And feel our noses chilling
And the chattering of
teeth?

What is this thing the weather?
I am hearing what you say.
It's the air and ground we see and
feel,
Every night and day.

Looking at the Weather

We can tell a lot about the weather just by looking outside. What do you think the weather is like in these pictures?

Draw what the weather is like where you are today.

  ISBN: 9781925726503

How do we Observe the Weather?

People observe the weather in many ways.

We use our **senses**. We look and listen and can feel the weather on our skin.

We use **technology.** Do you have any of these tools in your home?

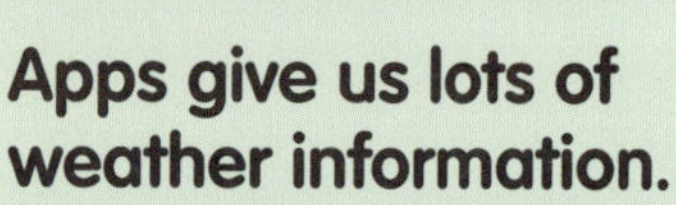

Apps give us lots of weather information.

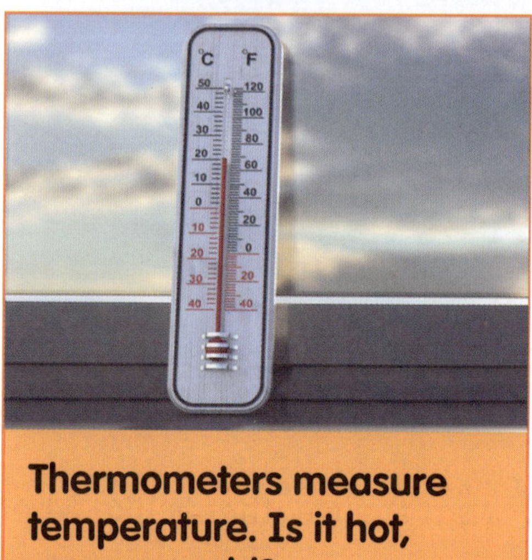

Thermometers measure temperature. Is it hot, warm or cold?

A rain gauge tells us how much rain fell.

Windsocks show where the wind is blowing from and how strongly.

TARGETING SCIENCE YEAR 1 © PASCAL PRESS ISBN: 9781925726503

Read this science rhyme to your child.

Temperature is measured.

Today, we have cool weather.

A **thermometer** tells me that.

I think I'll need a jumper.

A **rain gauge** will tell, dear,
The amount of rain that fell here.
Yesterday, 60 mls fell.
It sure has been a wet year!

A **windsock** will show
From where the winds blow.
A windsock also shows if the wind
Is blowing fast or slow.

Talk with Your Child

Have your child tell you how he or she knows what the weather is. For example, your child can see and feel rainy weather. Explain to your child that people use tools to tell us more about the weather. Look for a thermometer hanging in an outdoor space. Read the thermometer together with your child.

 ISBN: 9781925726503

How do we Observe the Weather?

Trace and read. Draw a line to match.

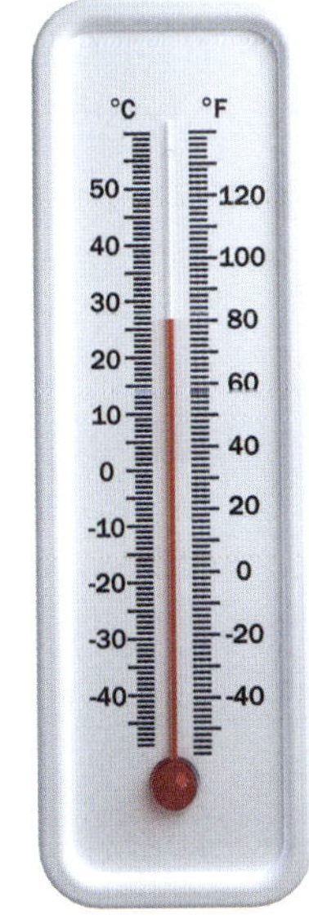

Ask an adult to help you use a tool or an App to observe the weather today.

TARGETING SCIENCE YEAR 1 © PASCAL PRESS ISBN: 9781925726503

Sharing Weather Observations

Weather scientists are called **meteorologists**. They use special machines and tools to observe the weather. Then they share this information with us often using symbols to help. Draw a line to match the symbol to a weather picture.

Do any of the pictures match your weather today?

Seasonal Changes

TARGETING SCIENCE YEAR 1 © PASCAL PRESS ISBN: 9781925726503

Sharing Weather Observations

Observe the weather for one week. Draw a symbol or symbols to show what the weather is like each day.

Symbols					
sunny	windy	cloudy	rainy	thunder	snowy

Sunday	Monday	Tuesday	Wednesday

Thursday	Friday	Saturday

Was the weather the same every day, or did it change?

____

____

Today's Weather

Together with your child, talk about the weather in your area. Does the weather change often? Help your child think about how the weather affects his or her life.

https://clickv.ie/w/M0gx

Use this QR code to access a video on this topic.

Observing the weather is important because it helps us choose what to wear and what we will do. What is the weather like today?
Circle the picture.

Draw yourself dressed to go outside today.

What are you doing today?

____

  ISBN: 9781925726503

Dressing for the Weather

Temperature has a big effect on what we wear. Draw a line to match.

Temperatures in Australia can be very different depending on where you live.

Does it get cold enough to wear gloves where you live?

Dressing for the Weather

Draw an **X** on the thing that does not belong.

1

2

3

Is the weather affecting your clothing choice today?

TARGETING SCIENCE YEAR 1 © PASCAL PRESS ISBN: 9781925726503

Weather and Our Activities

Weather affects our daily lives. Think about what happens when it is hot or when it is cold. Do you wear different clothes and do different things?

When we wake up and get dressed, we choose our clothes to suit the weather and our activities. If it is pouring with rain, we may have to change our plans for the day.

Look how rain might affect us.

What is the weather like for you today? ✔ the symbols.

Is the weather affecting your activities today at all?

Predicting Weather

Weather scientists use their knowledge to predict what the weather will be like in the week ahead. This is called a weather forecast. It can help us plan our activities.

This weather forecast is in a very cold part of Australia. Imagine you are going on a holiday to this place for a week. How would the weather forecast help you pack?

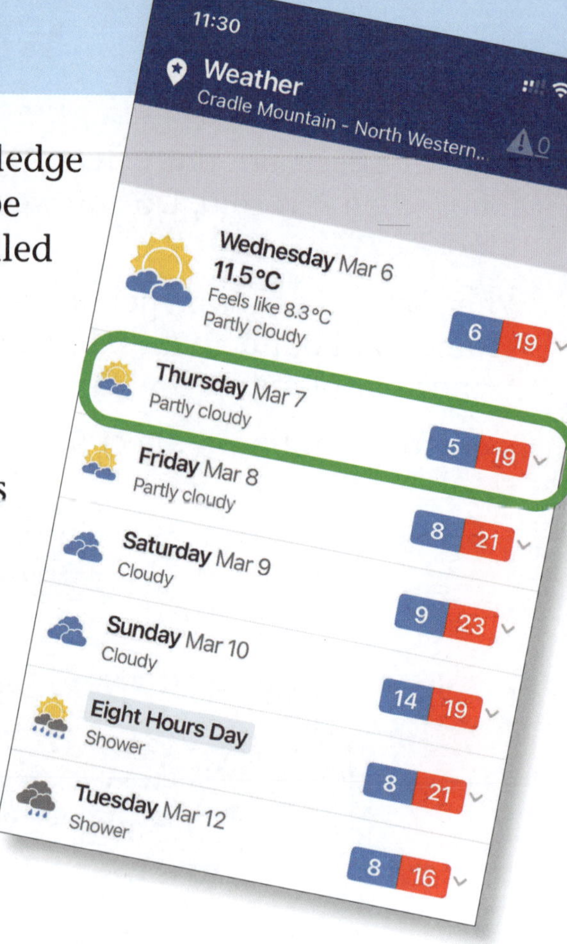

Circle the things you might wear or do on Thursday in this place.

TARGETING SCIENCE YEAR 1 © PASCAL PRESS ISBN: 9781925726503

Predicting Weather

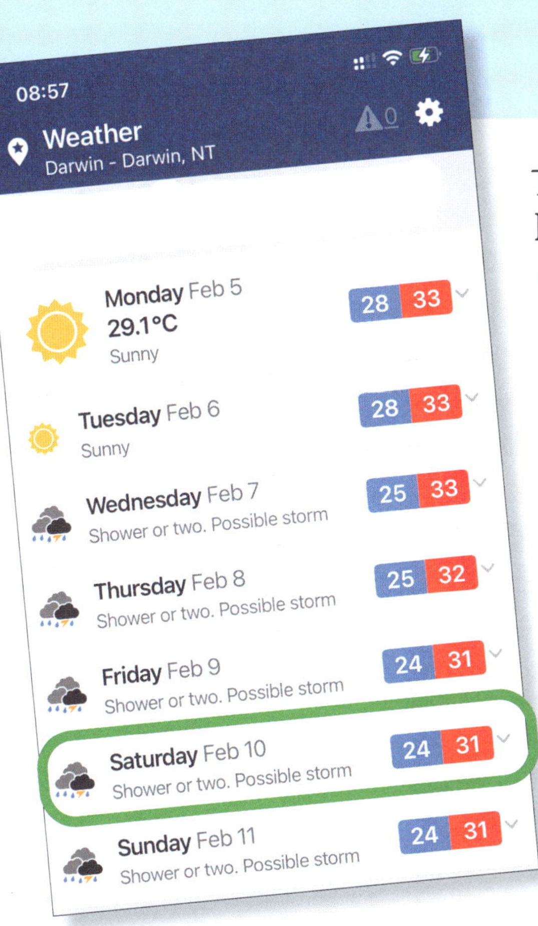

This weather forecast is in a very hot part of Australia. Imagine you are going on a holiday to this place for a week. How would the weather forecast help you pack?

Can you think of other times you might use a weather forecast?

Circle the things you might wear or do on Saturday.

TARGETING SCIENCE YEAR 1 © PASCAL PRESS ISBN: 9781925726503

What are Seasons?

https://clickv.ie/w/20gx

Use this QR code to access a video on this topic.

We know that the weather can change from day to day. But did you know that it also changes with the seasons?

Spring, summer, autumn, and winter are seasons. You may know that a season is a time of year that comes around every year. Each season has its own weather conditions and lasts around 3 months.

Seasons of the Year

spring
September October November

summer
December January February

autumn
March April May

winter
June July August

TARGETING SCIENCE YEAR 1 © PASCAL PRESS ISBN: 9781925726503

Observing the Seasons

Each **season** has special things that we can observe. We notice changes in the weather, and plants might look different as well. We change our clothing and activities to match what is happening. In Australia it often looks like this:

In spring, weather starts to warm up so we start to spend more time outside and take our warm clothes off during the day. Plants grow new branches, leaves and flowers.

In summer, the weather is warm or hot. We enjoy summer foods and activities and wear less clothes. Plants grow a lot and have lots of leaves.

The winter season can get very cold. In some places it can frost or snow. We wear warm clothes all the time and put blankets on our beds. We often spend less time outside. Plants grow very slowly or just rest.

Autumn - As the weather starts to cool the season changes to autumn. We start wearing some warmer clothes in the morning and nighttime. Plants grow more slowly and some start to drop their leaves.

Plants and seasons

Have you thought about the weather lately? What season do you think you are in?

Have you ever observed any changes in the plants in your area? All plants grow differently as the seasons change. Some we hardly notice as they seem to stay the same. Some change a lot. Their leaves may change colour and fall off. They may get flowers or fruit in some seasons. They may grow very quickly at some times of the year.

Draw some plants to show what season you think you are in at the moment.

Do the plants in your area change much with the seasons? How do they change?

____

____

TARGETING SCIENCE YEAR 1 © PASCAL PRESS ISBN: 9781925726503

All plants grow differently as the seasons change. Some we hardly notice. Some change a lot! Read this science rhyme to your child.

In **spring**, the trees
Will come alive.
New leaves and blossoms
Soon arrive.

In **summer** trees grow
Strong in the heat.
Plums and peaches
Taste so sweet!

Some **autumn** leaves turn
Deep red and brown.
Cold winds blow them
To the ground.

Winter comes.
Some trees are bare.
Some just rest
In the chilly air.

Winter passes.
No more cold rains.
A year of seasons
It's **spring** again.

 TARGETING SCIENCE YEAR 1 © PASCAL PRESS ISBN: 9781925726503

Animals and the Seasons

The seasons also affect many animals. Here are some ways animals behave in different seasons in Australia.

SPRING	SUMMER	AUTUMN	WINTER
Birds lay eggs and feed their young hatchlings.	Baby birds leave the nest and start to look after themselves. Adult birds might still help.	Most birds grow extra feathers ready for winter and huddle together for warmth.	Winter can be hard for birds with the cold and less food around. Some birds fly very long distances to warmer places. This is called migration.
When the rain starts to fall and it warms up, frogs start to become active. When there is plenty of rain, they lay their eggs.	The frog eggs hatch and the tadpoles eat lots of weeds and algae to grow and change as quickly as possible into frogs.	The adult frogs eat as much as they can to grow strong for the winter. There aren't as many insects now. Any tadpoles left will stay as tadpoles until the winter is over.	When it is cold, many frogs hide away and sleep in plants. Only a small number of insects live through winter. Some frogs hibernate in burrows underground.

Have you ever observed any of these things? Describe them to someone.

Animals and the Seasons

Whales also change their behaviour though the seasons.

In winter, whales arrive in warmer waters away from the very cold waters of Antarctica. They have their babies and feed them lots of milk to help them grow quickly.

In autumn, the whales swim back up to the warmer waters. They do this every year guided by the seasons. It is called migration.

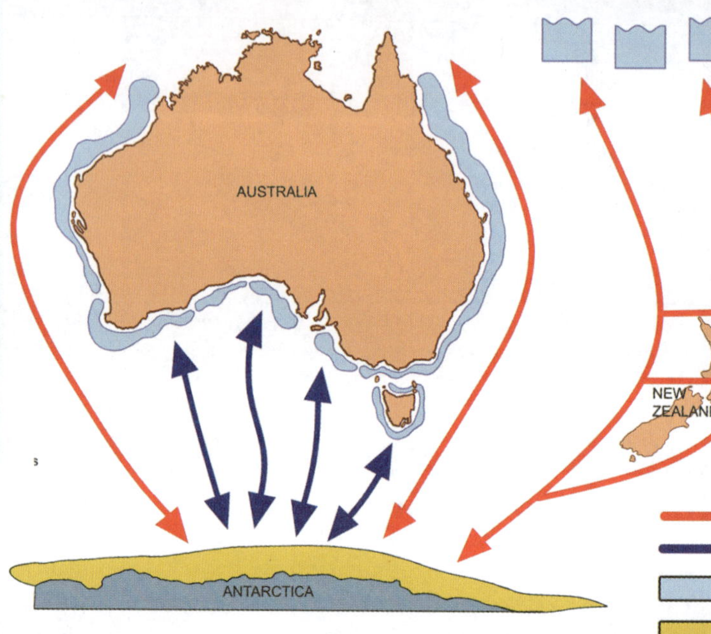

In spring, the whales start to swim back down to Antarctica. They swim in groups called pods to help the babies travel safely.

In summer, the whales eat lots of tiny plants and animals found in Antarctica at this time. As it starts to get cold again there is a lot less food so the whales start to swim back north again.

Have you been lucky enough to see a whale as it migrates?

TARGETING SCIENCE YEAR 1 © PASCAL PRESS ISBN: 9781925726503

Seasons Affect us Too

People are different to other animals. We have buildings to keep us warm or cool and clothes we can change to suit the weather. We can go to the shops and buy food no matter what the season. But seasons do still affect us. From summer to winter you may have noticed the biggest changes. Tick ✔ the things you have done in these seasons.

Summer	Winter

Draw something you do or use only in summer.	Draw something you do or use only in winter.

Seasonal Changes

Dress for Each Season

Draw a cross on the one that **doesn't** belong.

Seasonal Changes

TARGETING SCIENCE YEAR 1 © PASCAL PRESS ISBN: 9781925726503

First Nations Australians' Knowledge of Seasons

First Nations Australians have observed the seasons for over 60,000 years. They understand how Country changes with the seasons. This is how they have survived in Australia for so long.

Understanding the seasons means they always knew where to find food and water. They didn't have shops and taps they could just turn on. By observing the seasons, First Nations Australians knew when some plants would have fruits and berries they could safely eat and what season the fish and shellfish could be caught.

https://clickv.ie/w/Wwgx

Use this QR code to access a video on this topic.

Seasonal Changes

First Nations Australians' Knowledge of Seasons

By understanding the seasons, First Nations Australians knew when the flood waters would drop so they could travel to other areas for food and safety. Or when it was very dry, they knew where to find water to drink.

First Nations Australians use their knowledge of seasons to choose the best time for back burning so the fires stay small and make way for new plants to grow.

This prevents dangerous bush fires in the storm season which kill a lot of plants and animals.

 ISBN: 9781925726503

First Nations Australians' Knowledge of Seasons

First Nations Australians who live in the Torres Strait Islands have observed that the wind greatly affects their seasons. In Koki Kerker, the season that runs from January to April, they know there will be very strong winds and heavy rain called monsoons.

This means it is a dangerous time to be travelling by boat from one island to another and it is hard to catch fish to eat. But in Naiger Kerker, from October to December, the seas are calm because the wind is gentle. Travel and fishing in boats is safe.

Here is a drawing of the shape of Australia if you looked down from way up in space. Do you know where you live in Australia?

Australia is a very big country so the seasons can be quite different from one place to another. What season is it now and what is the weather like for you?

Seasonal Changes

Party time!

Imagine you are helping to plan a party. The weather forecast says it is going to be very hot.

What season do you think it would be?

____

Lots of people have been invited and because it is so hot it is going to be a pool party. But the weather app tells you there is going to be a big thunderstorm with lightning later in the afternoon. It is dangerous to swim when there is lightning.

How can you prepare for hot weather and the storm to keep everyone safe?

____

____

____

TARGETING SCIENCE YEAR 1 © PASCAL PRESS ISBN: 9781925726503

What are Objects?

Do you remember?

OBJECTS are things in our world which can be seen, held or touched. We use lots of them to live, play and work. Objects are not alive.

Some OBJECTS can be moved. How an object moves depends on factors like its size, shape and what it is made from.

Scientists ask questions and then investigate to find an answer. They work safely. They make predictions and careful observations. They compare what they predicted with what they observed.

What is a Force?

In this science unit we will be learning about something called **force** and what it can do to **objects**.

Force is how we make things move.

You are going to go to a door you can safely observe. It can be a cupboard door or a room door. Open and close the door a few times slowly and carefully observe how you are making the door move. Come back when you have made some observations.

What did you observe?

Now look at this drawer. What does the girl have to do to open the drawer or close it?

Did you think push and pull? We call these **forces.** What would happen if she pushed the drawer really hard?

Find other objects which you might safely push or pull to use them.

 ISBN: 9781925726503

https://clickv.ie/w/s0gx

Use this QR code to access a video on this topic.

Think of how you make things move. You can use your hands or your body to move something.

You can **push** something away from you.

You can **pull** something toward you.

TARGETING SCIENCE YEAR 1 © PASCAL PRESS ISBN: 9781925726503

Force Makes Things Move

A push or a pull is called a **force**. A force makes something move. A force can also stop something from moving.

You use more force to push or pull heavier things.

You use less force to push or pull lighter things.

TARGETING SCIENCE YEAR 1 © PASCAL PRESS ISBN: 9781925726503

Look at the picture. What kind of force is it?
Write either **pull** or **push**.

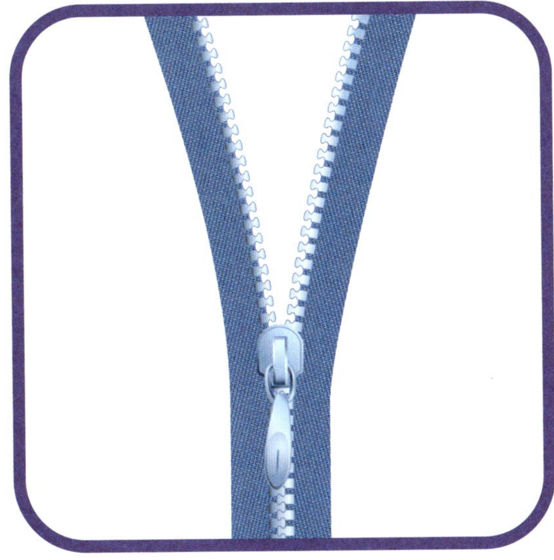

Look at the picture. Read the word.
Write the word in the sentence.

push

When you ________________ something, you move it away from you.

pull

When you ________________ something, you bring it closer to you.

force

A ________________ is something that makes an object move.

Think about this science question: Can we push and pull things with other parts of our body, not just our hands?

Do we use force when riding a bike?	Do we use force when riding a scooter?

When we ride a bike, we push down on the pedals with our feet to make the bike go forwards. Pretend you are riding your bike and watch what your legs do.	When we ride a scooter, we push on the ground to make the scooter roll forwards. Sometimes we need to stop the force of the push quickly, so that the scooter stops. How might we do this?

How do we use force when playing soccer?

When we are playing soccer, we can push and pull the ball with our feet and also with our knees, chest and head.

Making Things Move

We can also push or pull a ball in different directions.

What directions might the soccer player want to move the ball? Draw some arrows on the picture to show your ideas.

Think about how you can change the direction of a ball when you kick it. Explain your prediction to an adult.

Explain how we investigate kicking safely.

Find a ball you can kick. Try kicking it in slow motion in different directions on the ground (left, right, forwards) and in the air. Observe your body closely.

Explain how you change the ball's direction to an adult. Is this what you predicted?

Share another science question about changing directions in another sport or activity. This starter might help.

How do we use push and pull to change direction when we are ...

TARGETING SCIENCE YEAR 1 © PASCAL PRESS ISBN: 9781925726503

Changing the Direction of Objects

We can deliberately change the direction we push or pull a ball. By changing the way we stand and turning the body part we are using, we can make the ball go left, right, up or down.

Look at the pictures below. Predict where you think these people are aiming to push or pull the ball and where they wouldn't want the ball to go. Put a circle around the body part or parts that they are using to direct the ball.

Forces in Sport

Think about this science question: How does sport equipment help us push and pull?

Look at these pictures. Do you know what game they would be used in?

Sport equipment can make the ball travel faster and further than we can throw it with our hands. We can also choose the direction we push to make the ball go a particular way. But did you realise you use both **pull** and **push** to hit a ball with a bat or racquet?

First, we **pull** the racquet back ready to hit or push the ball.

Then we **push** the racquet forward and use lots of force to try and hit the ball hard in the direction we want.

Test different sporting equipment to feel the forces used when playing with bats and balls. Hitting a stationary ball on the ground allows you to explain the pulling back force, followed by the push force. Video the actions and replay it in slow motion to see the pull and push forces.

TARGETING SCIENCE YEAR 1 © PASCAL PRESS ISBN: 9781925726503

Read the diagram.

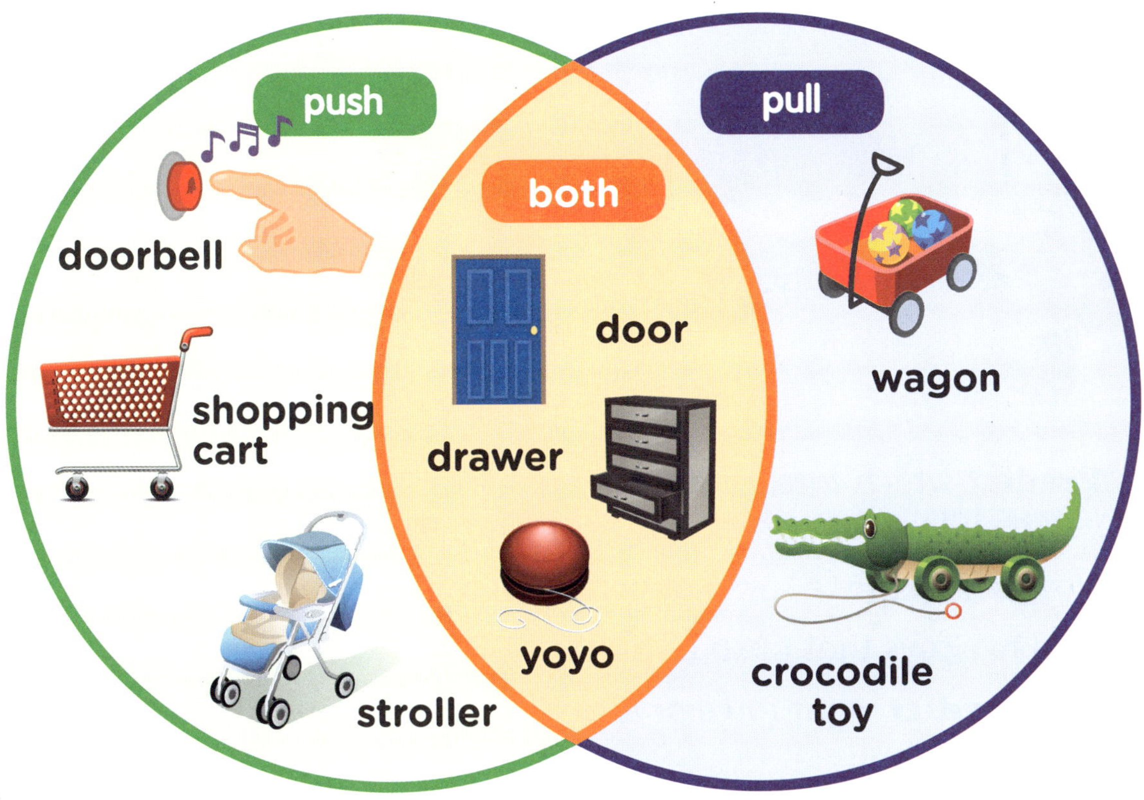

1. Name an object that moves with a push.

____

2. Name an object that moves with a pull.

____

3. Name an object that moves with both a push and a pull.

____

TARGETING SCIENCE YEAR 1 © PASCAL PRESS ISBN: 9781925726503

Forces and Toys

Think about this science question: **How do we use pushes and pulls when we play with toys?** Look at these toys.

Predict which force would you use with each toy. Do you think you could use both push and pull with these toys?

Now **investigate** some of your own toys.

Find two toys that push or pull.

First **predict** what force or forces you think you will use to make each work best.

Write this in the table. Now **test** each toy.

Record your observations in the table below.

Toy		Push ✓	Pull ✓	✓ Both push and pull
Toy 1	Predict			
	Observe			
Toy2	Predict			
	Observe			

Compare: Did your predictions match what you observed?

Findings: What can you say about pushing and pulling your toys?

___

___

TARGETING SCIENCE YEAR 1 © PASCAL PRESS ISBN: 9781925726503

Many Japanese people visit Australia every year. Some also live here. This is how we share things about our countries.

In Japan they also have toys that need pushes and pulls to play with them.

This is a toy called a kendama. The aim is to **pull** the ball to start it moving then **push** it gently on the other wooden parts so it bounces many times. People learn to do many tricks with the bouncing ball until they land it in one of the cups or on the spike.

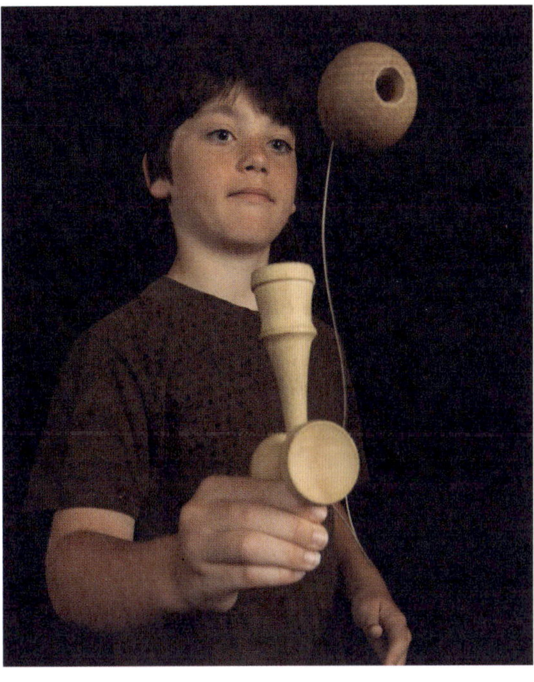

In Japan they also have a game called Daruma Otoshi. This game uses a Daruma doll in 5 pieces. The aim of the game is to use the small hammer to hit each of the coloured pieces out from under the head without the other pieces falling. You always hit the bottom piece until only the head remains. The hitting is a **push** just like with a racquet. Daruma dolls and their faces represent the importance of perseverance when challenged.

Investigating Forces and Shape

Remember when we do a science investigation, we follow steps in an order.

Science question: Can force change the shape of something?

Predict: Do you use force when playing with playdough? Yes No Unsure

Test and observe: Start with just the dough.

You will need:
- *some modelling dough*
- *a roller*
- *safe knife*
- *and shape cutters.*

Just using the dough alone, try pushing and pulling it in different ways. Can you see how your pushes and pulls change the shape? As you observe, try to say whether you are pushing or pulling and how it is changing the shape.

Try some of these ways.

 ISBN: 9781925726503

Investigating Forces and Shape

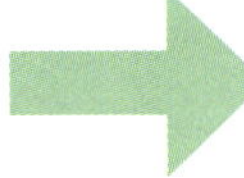

Now pick up the roller. Predict what the roller will do.

Was your prediction correct?

What force do you mostly use?

Do you have to use more or less force than when you were stretching the dough?

Now use the cutters and the knife. Carefully observe the forces you use.

What forces do you use?

____

How did you investigate safely?

____

Now push on the roller itself. Can you change the shape of the roller?

What might happen if you push too hard on a shape cutter?

____

Compare and think: What have you learned about force and changing shape?

____

Forces and Objects

Forces in Playgrounds

What do you think makes playgrounds fun? Playgrounds have many different types of equipment. Most of them need us to use force. Write if you would use **push** or **pull** to play on the equipment below.

____

____

____

____

Often we need both push and pull to have the best fun on play equipment.

The Dad is pulling the girl back

The girl is pushing hard to make the swing go forwards and high.

TARGETING SCIENCE YEAR 1 © PASCAL PRESS ISBN: 9781925726503

The children about to climb will also use both push and pull. Can you work out why?

The children will **push** up with their feet and legs and **pull** themselves up at the same time with their arms. It is a great way to build your muscles. This is one important reason that playgrounds are designed the way they are.

There are times when we need to use **strong** force and times when we need **gentle**. Look at the picture below.

What type of force is the mum using?

How strong would the force be?

What might happen if she used too much force?

We might even need to STOP the force. Share ideas about the pictures below.

You are on a swing. The boy has tripped and fallen right in front of your swing. How might you stop the force of your swing?

You are pushing a friend on a roundabout. She starts to get scared and wants to get off. What could you do?

Direction of Forces

We can show a force on a drawing using arrows. Have a try yourself.

The man is pushing the mower forwards. The arrow shows us this. 	The man is pushing the shopping trolley forwards. Draw the arrow. 
The door can be pushed and pulled. The arrows show us this. 	The drawer can be pushed and pulled. Draw the arrows. 
The ball has been pushed forward to the bat. The bat pushes and changes the direction of the ball. 	The ball has been pushed forward to the bat. Draw how the ball will change direction. 

TARGETING SCIENCE YEAR 1 © PASCAL PRESS ISBN: 9781925726503

What You Need

- chair
- book
- pot or pan
- paper clip
- toy with wheels

What You Do

1. Collect each of the objects listed above.
2. Predict the forces you use to move each object.
3. Name any safety things for the investigation.
4. Push or pull each object and observe how much force is needed.
5. Write about what happened in the chart on the next page.

How Much Force?

Item	Push or Pull	What happened? Did you have to use a lot of force? Why?
	pulled	The wagon moved forward; I did not have to use a lot of force.

Remember your predictions. Did anything surprise you as you investigated?

TARGETING SCIENCE YEAR 1 © PASCAL PRESS ISBN: 9781925726503

Write a sentence about each picture.

Do you need to use more force to push a car or a bike? Explain your answer.

Designing an Obstacle Course

Have you ever played on an obstacle course?

An obstacle course is made with equipment and objects laid in a line that you follow. The equipment gives you different body challenges to do as you move along the course. There are lots of types, even ones in water.

What is a science question you could ask about one of these playgrounds?

____

Forces and Objects

TARGETING SCIENCE YEAR 1 © PASCAL PRESS ISBN: 9781925726503

Designing an Obstacle Course

Design Challenge: How can we design an obstacle course to include push and pull activities?

You are going to design an obstacle course using your science knowledge of push and pull. You can draw it, make it from playdough or plasticine, or use equipment around your home to set one up. It can be inside or outside but must have at least **3 push** and **3 pull** activities. It can have a start and a finish or be in a circle. Make sure your activities are **safe**.

1. Take time to plan before making it. You will need to be able to tell someone about the 3 pushes and 3 pulls. It might include things like:

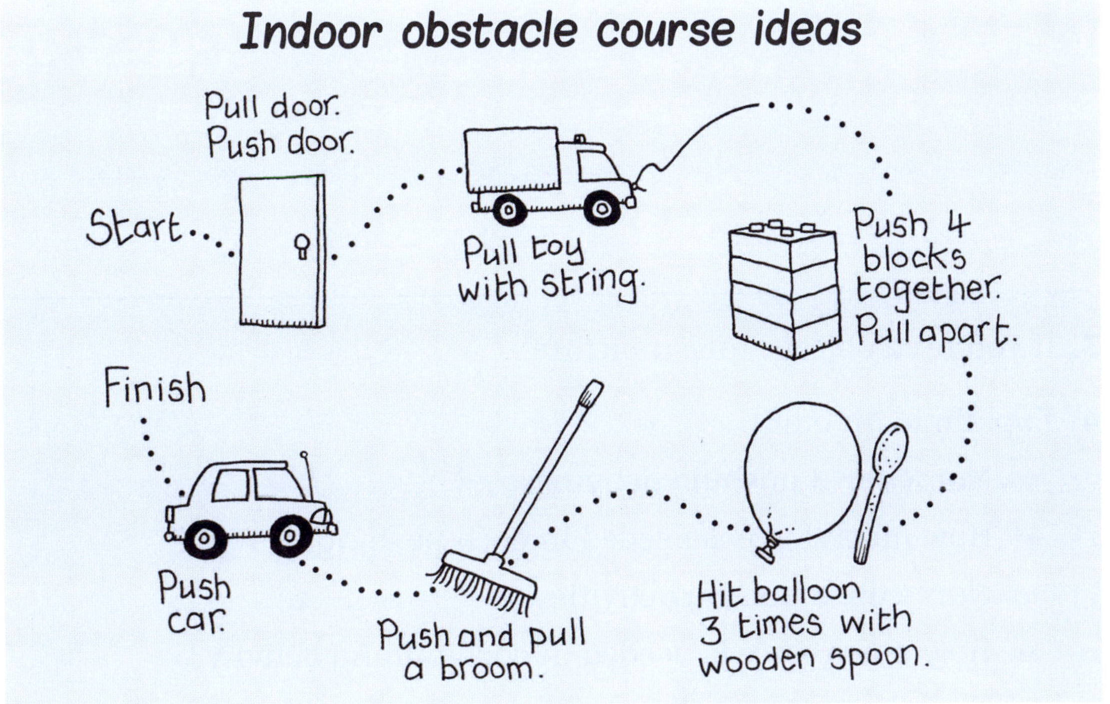

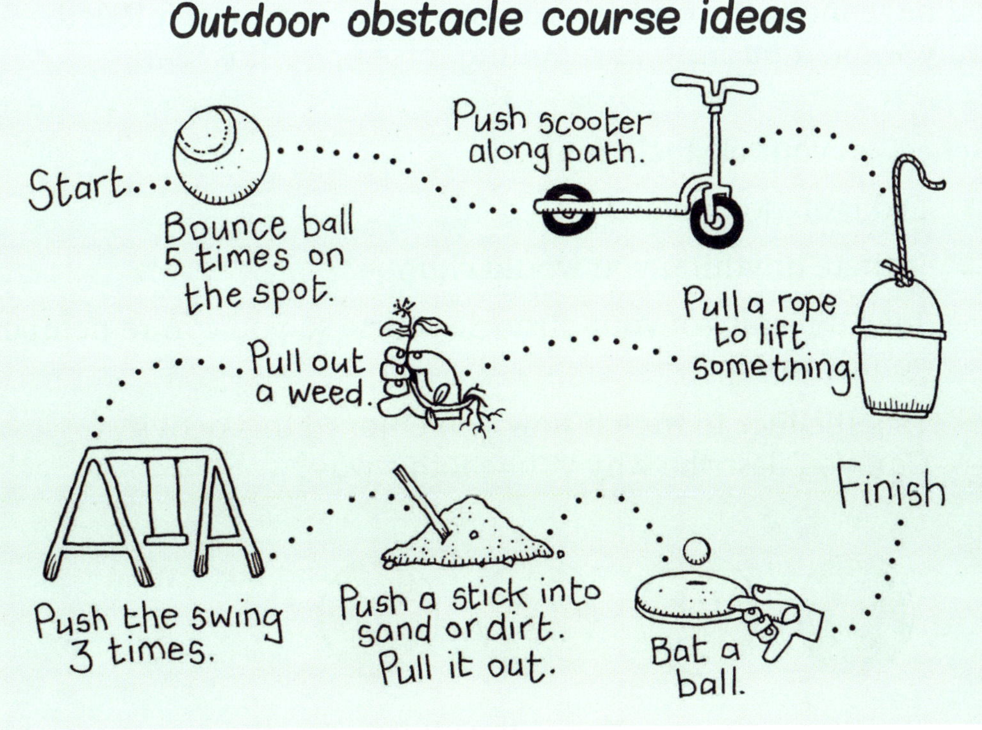

Designing an Obstacle Course

2. Draw or set up your obstacle course.

3. If you used objects, take a picture …

4. Explain to an adult.

- ✔ Name the 3 pushing activities.
- ✔ How much force is needed in each pushing activity?
- ✔ Name the 3 pulling activities.
- ✔ How much force is needed in each pulling activity?
- ✔ Describe anything they need to know about keeping safe.

If you have made the obstacle course with objects, test it together. Identify when a pushing or a pulling force or both is used.

5. Reflect on your obstacle course.

- ✔ Did it all go to plan?
- ✔ Is there anything you would change?
- ✔ Take a picture of your obstacle course so you can remember your learning.
- ✔ Tick the face to match how you enjoyed this activity. Can you describe why you feel this way?

Page 4

puppy: home; seedling: garden bed; fish: ocean; platypus: creek

Page 9

tick: top left and top right; dot: all the rest

Page 10

orchard: fruit tree; sea: starfish; bushlands: echidna

Page 11

nest: birds; pond: fish, frog; tree: owl, possum

Page 12

The plants needs sunshine, good soil and water. If their needs are not met, they will not thrive and will eventually die.

Page 13

flowers: birds and bee; fruit and vegetables: child; grass: cow

Page 16:

mangroves: crab; boat: fish; bush: koala

Page 17

red: glass of water, clothes, salad, bed

blue: berry bush, nest, insect, tree, pond

Page 18

red: rain, garden bed, soil

blue: fish, reef, coral

Page 19

food, water, air, sunlight, shelter

observation; vet, botanists, farmers, breeders, gardener, pet owners

Page 20

They would become less healthy and would need to move to another location.

Page 23

sunny, snowing, windy and raining, hot and dry

Page 27

rain: rain gauge; wind: wind sock; heat: thermometer

Page 31

hot: thongs, shorts, board shorts; cold: mittens, beanie

Page 32

1 sunhat; 2 shorts; 3 beanie

Page 34

Pack clothes for hot weather as well as umbrellas and raincoats.

shorts, thongs, beach

Page 35

Pack clothes for hot weather as well as umbrellas and raincoats.

gumboots

Page 44

winter: beach chair; spring: autumn rain; summer: snow gear; autumn: beach gear

Page 48

summer have sunscreen, shade, and cold drinks, when it is hot. Watch the sky, and make sure everyone is inside before the storm hits.

Page 53

pull: zipper, tissue; push: swing, toy

Page 58

tennis, cricket, catch and throw, baseball or softball

Page 59

push: doorbell, shopping cart, stroller; pull: wagon, crocodile toy; both: door, drawer, yoyo

Page 60

Findings: e.g. both my toys worked better with a push. I can push and pull both toys. My pull toy has a handle/string.

Page 64

push, pull, push, pull

Page 69

The man is pushing the car, it takes a lot of effort.

The girl is pulling the wheeled suitcase, it does not take much effort.

You need more force to push a car than a bike because the car is much heavier.

Page 70

For example; What materials can be used in an obstacle course? How do they build a water course? How do they keep safe? How do the pool noodles stay up?